DYNAMIC POSITIONING CONTROL SYSTEMS

*A Technical Companion to
'Anchors are for Wankers' and
'Poetic Lessons in Life'*

By

Doug Phillips

aka DP Doug

I'd like to dedicate this book to my grandchildren: Tia Wayland, Lucas Moore, Leila Adams, and Marnie Moore, who have brought a new light into my life.

CONTENTS

ACKNOWLEDGMENTS

Brian Haycock who stuck with me no matter how off beat I got, he also kindly provided the majority of the illustrations in this book. Some of the other DP control system geeks I have had the pleasure to work with and learn from – Richard Bond, Bruce Kauffman, Tony Wilkins, Alan Davis, Holger Rokberg, and Nils Albert Jenssen.

Having completed the manuscript for this book. I had some last-minute reservations about publishing a technical subject in such a different way. Then I happened upon the BBC's 'The Entire Universe', which is a one-hour musical comedy come pantomime, with Eric Idle and Professor Brian Cox. Along with Professor Stephen Hawkins singing in the closing number. I thought that if Cox and Hawkins were up for treating their subject in such fun, why shouldn't my subject be done in verse? My mind was made up – publish and get over it.

BACKGROUND

The 50 years in engineering

Leads me to present this all logically

But the poet wants to just go with the flow

That's the curse and dichotomy

Of trying to be a poet engineer

Where it's the rhyme that tends to steer.

If you don't know much about dynamic positioning

It would be a good idea

Before you start reading

To look it up on Wikipedia

Like the weight of the Albatross
Around that Ancient Mariner's neck
My DPCS knowledge hangs like his curse
But I don't want it to get lost
So I thought, 'what the heck'?
I'll knock it out in verse

That way I get to pass it on
And at the same time have some fun
Rather than mired in reams of prose
For which I don't have the time, or patience I
suppose.

Song lyrics often have rhyme but no reason
Prose has reason but no rhyme
But with poetry you get reason and rhyme
One bullet point at a time.

So let's let that albatross fly free
In the following poetry
That will benefit hopefully
Both you and me

If you want to chip in

I can include it in the next edition

Like open software, but open poetry

Don't sweat the rhyme, that can come from me.

INTRODUCTION

Positioning a ship dynamically

Is keeping a ship still

By only using thrusters

So it can maybe dive or drill

Known as 'DP' - Dynamic Positioning

It covers many an engineering discipline

All of which it's impossible totally to know

After a career life time of DP, I still have a way to go

DP Experts, DP Consultants, DP Operators

Who don't know what they don't know

But in the land of the blind, the one eyed DP 'expert'
is king
Admit they don't know something – what will the
blind be thinking?

The Venn diagram
Tries to summarize the problem
Even in what we thought we'd known for long
There are places where we could be wrong.

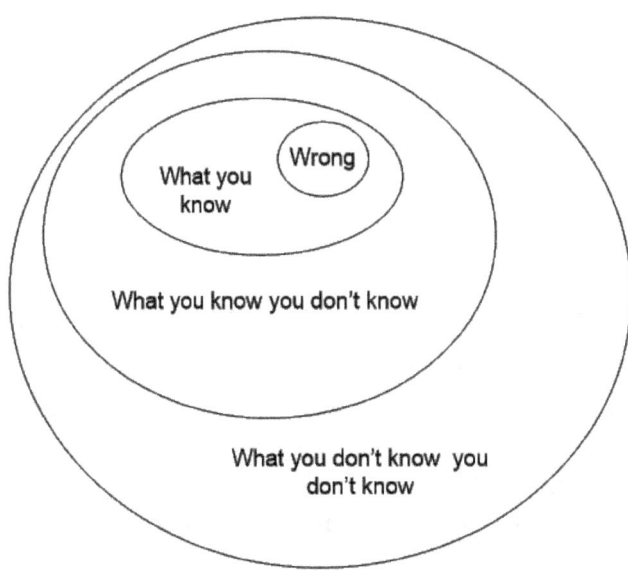

The word anchors
Are not in our vocabulary
Anchors are so passé
They are just not necessary

There's too much subsea junk
Setting up takes way too long
Or the water's too deep
Or you need to move, if a hurricane comes along

To DP or not to DP?
Nowadays most times, one can't use anchors
There's often no other choice you see
So one puts one's trust in thrusters

The earth's spin, sun and moon create
Current, wind and wave forces that
To which the ship will react
But when it moves off, the DP puts it back
Using the thrusters, these forces to counter act.

It's not just control of position
Rec Stanbury once explained to me

All we need to be really doing

Is keep it at zero velocity

When something that big starts to move

There's no easy way of stopping it

More and more quickly position you'll lose

And everything soon 'turns to rat shit'

A ship has six degrees of freedom

So let's examine them

Only three we control, no more

Surge, Sway, Yaw

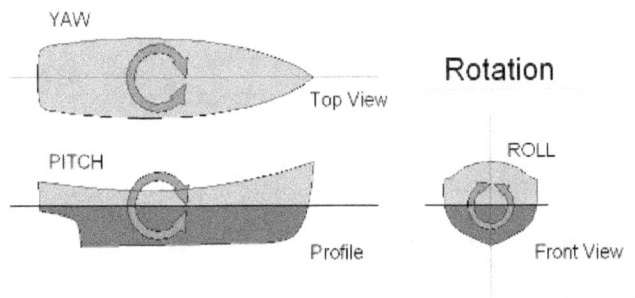

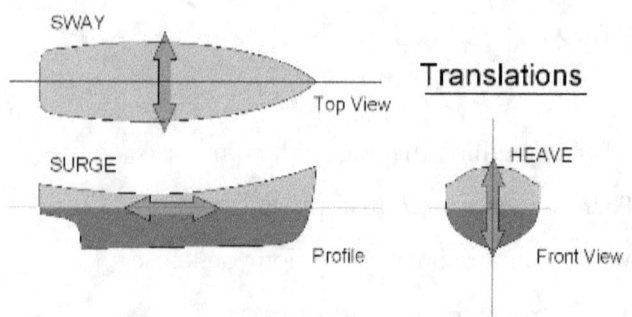

Also known as X, Y, N

Fore Aft, Port Starboard, heading

The others; roll, pitch, and heave

For control purposes we can leave

Heave is for indication only

Roll and pitch is measured timely

For the DP to remove their noise

So, thruster induced roll and pitch, we can avoid

Thrusters can be used to reduce roll

But that's the exception, not the rule

Especially if fitted with fast acting cycloidal propellers

Which for DP are inefficient and not that popular.

The diagram shows it all complete

The DP closed loop control system

It's mainly six blocks in the grey that we'll treat

And deal with each of them.

(Ref MTS DP Conference Proceedings – Introduction to
Kalman Filters by Olivier Cadet)

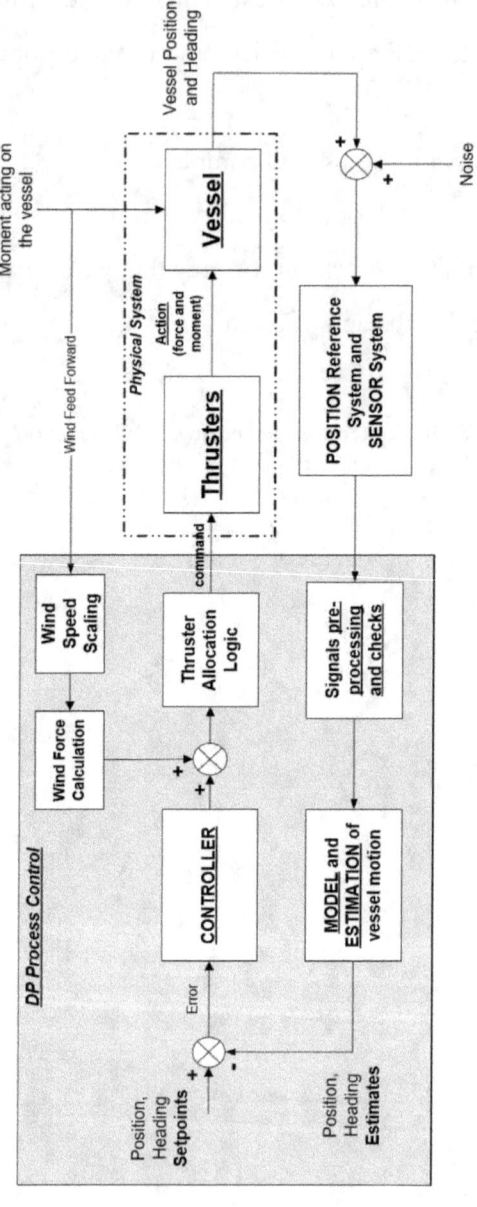

POSITION AND HEADING FEEDBACK

Any closed loop control system

Needs a reliable feedback

And a DP position and heading control system

Is no exception to that

There are many choices

It's a shame that none of them

Are totally perfect

So some we have to reject

Of the many systems
Three are needed
For position feedback
So redundancy it won't lack

At least two of the three
On different principles need to be
And none should share
A common way of failure

For what's known as PRS or PME
(That's Position Reference Systems
And Position Measuring Equipment
To you and me)
Plus measurement of heading, needs three

The advantage of three
Is that in a vote you get a majority
Which can thereby reject
Anything behaving erroneously

All are not created equally

Take just a basic three

A taut wire drifts with the tide

But is mainly steady, albeit with catenary caused

inaccuracy

It does have good repeatability.

Acoustics are accurate mostly

But they may act nervously

In the face of noise or aeration

More jumpy means even more noise and aeration

The vicious circle spirals to a loss of station

DGNSS may drift off

Maybe your differential provider cuts you off

Or it fails the face of sun spot activity

Or freezes and acts seemingly 'perfectly'

I have skipped through these three

Fairly quickly

As this poem about DPCS

Not the PRS, or PME

It doesn't just need to know where you are
So it can calculate your position error
It's also needs to reject measurements of where you're
not
To keep you on the desired spot.

This we will come back to later
When we cover the Kalman filter
But here also need to mention
PRS need to be suitable for the water depth
And the ship's industrial mission

A suitable PRS needs good repeatability
Rather than pin point accuracy
Unless you need an exact point on the seabed
Then most times there will be a separate survey
system instead.

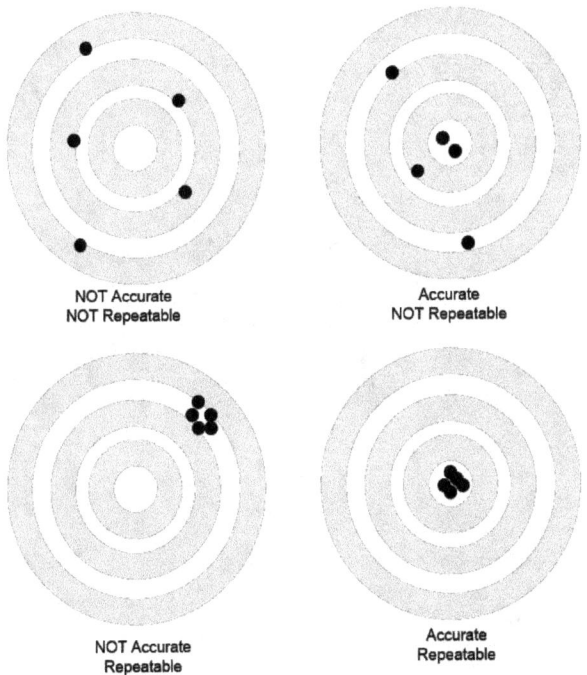

NOT Accurate
NOT Repeatable

Accurate
NOT Repeatable

NOT Accurate
Repeatable

Accurate
Repeatable

THREE TERM CONTROLLER

So position error and thrusters

That's what it's about

Ship's position and heading error goes in

To the controllers for each X, Y, N

(Fore Aft, Port Starboard, Turning)

Thrust for each comes out

Based on three terms of controller action

One to the error is proportional

Two opposes the speed, aka differential

The third action is integral

For you a simple example
Of what three term control
Is all about
Think how you'd steer a boat.

The amount of rudder you apply to stay on course
Is proportional to how far you are off course
Which is Kp and not enough alone of course
'Cos you'll tend to overshoot your desired course

So you have to apply a little counter rudder
Before you get to the exact course, you need to slow
her
That is differential term Kd
Soon you'll get it just right you'll see

Too much Kp not enough Kd
Control will be unstable and skittish
Too much Kd not enough Kp
Control will be too slow and sluggish

If a side current or wind either way
Always forcing you off heading one way

The rudder will have to have a bias to one side
This is the third term integral action – Ki

These actions should remind a lot
Of those on an auto pilot
'Rudder gain', is action proportional
'Counter rudder,' is action differential
'Weather' is the longer-term integral.

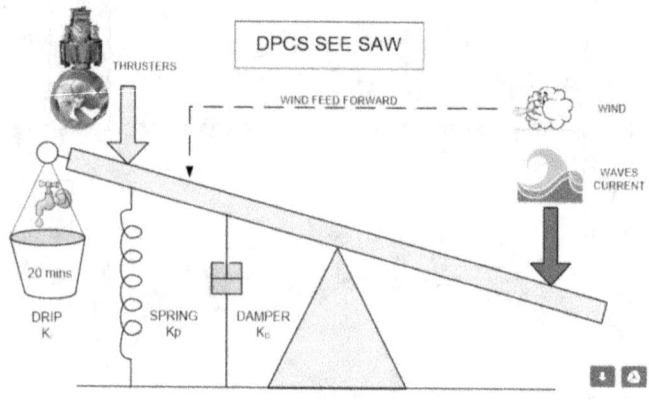

This see saw gives a good analogy
At one end wind, waves and current
At the other the DP needs to get the thrusters
To balance the environment

The spring hold things steady initially

Spring constant Kp

Until the drips in the bucket

Fill the bucket enough to relieve it

That's the integral term Ki

The other is damping Kd

Just like the dampers on your car

That takes the bumps out to give you a smooth ride

The wind can be either side

But if wind comp you don't select

The wind force that's on right

Ends up in the integral bucket

The waves and current

Change slowly over hours

So the dripping integral tap

Can easily track that

But the wind can be quite gusty

Or even shifty

Changing potentially very swiftly

And should be compensated for directly

To cope with all that
The DP takes measurements from wind sensors
Along with a drag model
Calculates the resultant forces

These are 'fed forward'
Onto the controller's output X Y N
To apply thrust before
Any position or heading loss, that's wind driven

Some care needs to be taken
With wind compensation
When it's selected and deselected to ensure
A nice smooth transition.

This involves subtraction or addition
From the integral drip bucket
To ensure that smooth transition.
The amount needs to be correct

The integral bucket
Here's a clever bit
Is the steady forces on the ship that prevail
So it can be used if all PRS should fail

This is known as 'Model Control'

(AKA Dead Reckoning)

And for short periods is useful

Until either another PRS is made available

Or the operator takes over in manual.

THRUSTER ALLOCATION LOGIC

Thruster Allocation Logic, 'TAL'

Is next what this is all about

X Y force and N turning go into the TAL

Then individual thruster demands come out

The math can be complicated

You could hear terms like 'optimized'

And 'iteration', 'Lagrangian'

And all types of 'thruster interaction'

Generators power the thrusters
That's also what it's all about
The generators put the power in
The thrusters put the power out

The generator power that's available
The DP has to stay within
Such that after a worst case failure
Other generators or thrusters don't end up limiting or
tripping.

The thrusters provide the thrust
Just enough to keep a vessel still
Always remember this
As Howard Shatto put it
'You can't drill if it ain't still'

The Star Pisces back in the seventies
Had the simplest TAL of all
But it demonstrates the point easy
A practical example
Of why you did simultaneous equations in school.

A simple thruster transform/thruster calculation

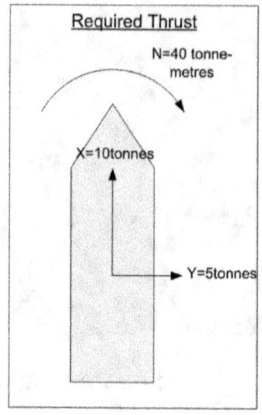

Required Thrust

N=40 tonne-metres

X=10tonnes

Y=5tonnes

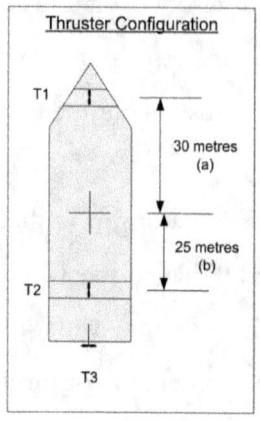

Thruster Configuration

T1

30 metres
(a)

25 metres
(b)

T2

T3

$T3 = X = 10\text{tonnes}$

$Y = T1+T2$

$N = a \times T1 - b \times T2$

$T1 = \dfrac{N + b \times Y}{a + b} = \dfrac{40 + (25 \times 5)}{30 + 25} = 3 \text{ tonnes}$

$T2 = \dfrac{a \times Y - N}{a + b} = \dfrac{(30 \times 5) - 40}{(30 + 25)} = 2 \text{ tonnes}$

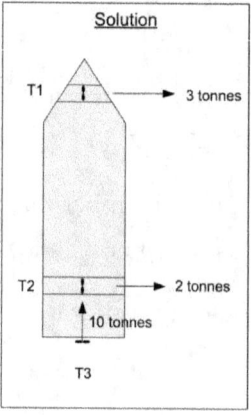

Solution

T1 → 3 tonnes

T2 → 2 tonnes

10 tonnes

T3

THE MODEL AND ESTIMATION OF VESSEL POSITION

Used to be done down and dirty

Many a true word spoken as a joke

'Just tune for maximum smoke'

But like that, the thrusters soon broke

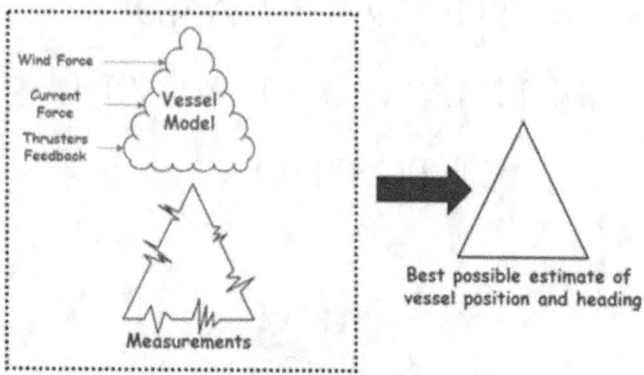

Best possible estimate of
vessel position and heading

But now we have a model

Details soon to follow

Which the fluffy triangle represents

Which removes the spikes on the PRS

Because the model shows them to be, impossible

movements

As so nicely illustrated in the diagram

Approximations in the model

Make its predictions 'wooly' but not noisy

Unlike the noise on the measurement triangle

The model actually learns

Using such terms

As innovation and correction
By comparing the actual with its prediction

How do we achieve this so smoothly
We'll need to first go back centuries
To Sir Isaac Newton
And his second law of motion

He came up with this equation
During his two-year bubonic plague isolation
Force equals mass times acceleration
Aka Newton's Second Law of Motion

Once we've worked out **'F'** and we know the **'m'**
We can then calculate **'a'**
From which we can get velocity
And from that an estimate of position
That's smooth, without noise, and jumpy transitions

This is simplified of course
But it'll serve its purpose
Einstein later got it exactly right
But we aren't going near the speed of light

For a ship we also have to note

There's added water mass

Dragged around like a moat

Plus the water drag that's on anything afloat

But we have enough info to give it a shout

Another clever man

Was Rudolf Kalman

Who came up with the methodology

That DPCSs along with a Newtonian model use today

Enough to say

It's how to mix actual measurements

In an optimum way

With that from an updating model

And without a conventional lag filter's

Undesirable delay

By giving a smooth estimate of velocity

It gives the 'laid-back' control we get today

The main issue with Kalman

Is its mathematical complication

Which makes explaining it
As you've likely gathered, rather problematic

It should be a Kalman Bucy
But Bucy got dropped along the way
Or linear quadratic estimation
We'll just stick with, 'Kalman Filter' notation

But MTS has the math answers explained right
Find the paper on www.dynamic-positioning.com
website
There you'll find one of the best yet
Introduction to Kalman Filters by Olivier Cadet

There is one issue though not to forget
The model may use actual thruster feedbacks
Which is OK, until one goes wrong one day
Then what's going on, the model has to guess
Someone needs to stop the faulty one
Before all three DPCS's, get in a total mess.

POSITION REFERENCE PROCESSING

The design needs to care

Not just calculating where you are

But also rejecting on the dot

Anything indicating where you're not

The problem therefore comes as three

First –is where you want to be

Second – is where you are now about

Third – is deciding where you are not

Then the math is in north and east

You'll need one heading measurement at least

So in heading you can move freely

And the errors in ship's axes track truly

Loss of all gyros

You'd think was hard to do

It loses not only heading control

But all position control too

This has been achieved

By all having common connection

To a backup 24V that spiked

Or GPS jump to all gyros, set to auto speed and

latitude correction

Having cleaned the errors up, the easiest to do

Is to put them in a priority queue

This can be an operator set priority

Or maybe use the average for simplicity

But these will give jumps

When one PRS gets dumped

Most DP vendors will agree
Better to move between smoothly.

Maybe even work out, how noisy each be
And weight them accordingly
But this gives some extra trouble
If they don't come in every second on the double

Then you'll need logic to detect and reject
If one of them seems to be perfect
Or worse freezes or drifts slowly
You can see, none of this is easy
For this we have the median test

Inertial (INS) presents an interesting possibility
It gets position from acceleration via velocity
(Just like the DP model you can see)
Using the second law of motion
Force = mass times acceleration

It then also gets ship position
From measured acceleration
Via velocity

But for some this is just too costly

And because it will quickly drift

So it also isn't perfect

Another way to look at this is

That the model is another PRS

That's just like an internal INS

But it costs a lot less

And gives you dead reckoning none the less

'If only there were one perfect PRS'

Laments the designers of DPCS

'We thought we'd found it with DGNSS

But we always still need other PRS'.

NOISE

Each **PRS** has its characteristic noise to remove

That its weighting and the model smooth

But on the ocean, there's always motion

Because of the waves action

Causing a ship to slowly drift

As well as heave, roll and pitch.

Wave action has two components you see

Both high and low frequency

High is where the ship moves around in a wave

But pretty much goes back to where it came

But each time there's a slight shift

As the wave motion forces aren't quite symmetric

This gives the slow low frequency wave drift

And the control loop easily deals with it

The high frequency motion forces are huge

And to the thrusters mustn't get through

So a high frequency model is used

To be sure that from each PRS, the high frequency

noise is removed.

Heave is displayed only for info

For say heli ops and drill floor

But roll and pitch there needs to be more

As most PRS need their measurements corrected for

Acoustics need to do it directly

It needs to be done with no latency

Here the errors would be proportional to water depth

Except in long baseline, where it's just the transceiver

depth

Taut wire doesn't need it at all

If the taut wire angles are to the vertical

But if measured to the ship's hull

It'll need correction for pitch and roll

The rest are done in the DPCS software

While others have their own sensors

Some at the wheelhouse top height are not worth the bother

Others at the top of a drill derrick are a different matter

Before Kalman filters were adopted

Wave motion removal was by notch filtering

But these put a significant delay in the feedback position

Moving closer to instability in phase margin.

DP CONTROL SYSTEM (DPCS) REDUNDANCY

DPCS with triple redundancy

Is more the norm now seems to be

But each of the processors three

Have identical software and 'personality'

Even our DPCS let's call it Triplex

(just to jest and Kongsberg a little vex ;o))

Would surely confess

That under certain duress

All three can crash, leaving us in a mess

There's always the possibility
Regardless of software maturity
Of a failure of all three

That's why we have an Independent Joystick
A-K-A I-J-S
And manual thruster controls
A-K-A MTC

Be especially wary of the dangers
Of those software changers
Management of change – M O C
Needs to be taken very seriously

NASA's motto is 'Failure is not an option'
But how well did that work for them?
For DP, failure is an option
But failing to plan for it, is not one

If worst case failure is enough shocking
All the redundancy groups on a diagram Venn
Even with a triplex, things are overlapping
All the remaining loads will be doubling

If something else additional is tripping

Things will be even more troubling

In most failure analysis and trials this is missed

But worse case the ship off position will drift

Used to be alarms and controls

Were treated separately

Now we put everything on a dual network

And relies instead on redundancy

Of course it doesn't always work

When it totally fails

The vendor will be driven near berserk

With the ship spending weeks out of work

Our only mitigation to it

Is the independent joystick

So test it make sure

You can still select and use it

After both networks' failure

Then for manual thruster controls, repeat it.

SUMMARY

By now your likely half crazed
But at least take that away today
But finally all in all
We should be amazed
That it works at all

Let's all get coveralls or tee shirts
That on the front should go
'Anchors are for Wankers'
With DP Doug's road sign logo

And on the back saying:

'Put your Trust in Thrust'

'Moorings are so Boring'

"POSMOOR A T or A T A is so passé"

The fifty years an engineer

Weren't all in vain

The Engineering poet

Has struck again

ADDENDUM - 'DP DOUG'S'
TOP 20 DPCS TIPS

Here are examples that to be aware.

If you have your own favorites let us all know.

The airline industry sees it as positive to share

Marine and offshore much less so

Medical profession – how do they do ?

On that you need to read the 'Checklist Manifesto' .

In no particular order let's see how you fair

Read on and any shortcomings may you beware.

1. Always be open to learning how the DPCS works, even though this might be over something you thought you knew already. Stay curious and 'floreant denditae' (may your brain cells flourish).

2. If there's a software change satisfy yourself that it is fully tested under a documented management of change. There should be no possibility of it causing unexpected issues elsewhere in the DPCS. Remember that Chernobyl was a result of trying to introduce improved safety features!

3. Doesn't matter how many ways the power system can be split by opening bus ties, they still have a system that can defeat the redundancy concept for the total DP system. That is the DPCS itself which in the event of failure of one section will immediately increase the load on the remaining power plant. If it doesn't have the capacity or ability to take the step load, then position will be lost. So be sure that the DGs and thrusters can give 100% if you need it. Then if the worst case failure occurs and the DPCS

doubles the load on the remaining ones nothing else should trip.

4. Don't be over confident in the fact the vessel has a DP notation from class – it is equipment class only. It doesn't mean you have the right equipment, modes and features to achieve a particular industrial mission.

5. Know and practice how to handle the vessel in IJS and MTC

6. Know and practice how to select IJS and MTC from DP control in an emergency.

7. Be sure someone has tested that IJS and MTC are still functional with both networks off.

8. On the DP joystick make sure full power is already selected in case you need full power to pull off.

9. Gyros with GPS inputs for latitude and speed correction, should be set to 'manual'. Otherwise a faulty GPS can affect all three gyros. Also make sure gyros are not on the same 24 Volt back up supply as a glitch on that has caused all three to fail.

10. Of external interfaces such as draft sensor, tension reading, fire monitors etc. be wary. Use manual inputs instead if the interface can't be shown to be engineered, checked, and bounded correctly.

11. Be wary of mixing absolute PRS (e.g. DGNSS, acoustic, taut wire, etc.) with relative PRS (laser and radar based).

12. Obvious though it may seem – a relative system with a target on a fixed installation is an absolute system.

13. If you are following an ROV without a reaction radius then this is relative.

14. Use alternate center of rotation (COR) with care, if possible use it near the center of gravity (C of G). Alternate COR is basically making position and heading changes at the same time – this is not allowed if diving according to IMCA.

15. Whenever possible check the consequence analysis warning is correct. Don't however rely on it totally.

16. Do use the median check when you have three PRS or more. This will reject seemingly perfect

PRS that would soon monopolize the PRS weightings.

17. Be careful with vessels with a single stern thruster powered by a shaft generator from a main prop. Failure of one main engine may leave a vessel with one bow tunnel, one main prop and one rudder. This needs sufficient ahead environment to dissipate any unwanted ahead thrust from the use of the rudder.

18. Always use an agreed ASOG/Decision Support Tool for the activity you are embarking upon. Make sure it or your other checklists cover each of these tips.

19. Send me your tips to dougphillips@dpexpertise.com

20. Finally –

DP Fail- ligator

Five monkeys swinging in the tree
Teasing Mr Alligator - 'can't catch me'
Mr Alligator sneaks up on them – quietly
Snaps that monkey right out of that tree

Four monkeys, three, and so on
So goes our granddaughter's song
With all the actions that go along
Until of course all the monkeys are gone

Many DPOs sitting on DP
Teasing 'DP Fail-ligator you can't catch me'
'I've got a certificate for class 2 or 3'
DP Fail-ligator creeps up on them quietly

Knocks that ship right across the sea
Shocks everyone out of their complacency
Don't get caught out like those monkeys
Expect DP Fail-ligator constantly.

DP GLOSSARY, ACRONYMS
AND ABBREVIATIONS

These include many that aren't in this book but you
might find them usefull if you read much else about
DP elsewhere.

AA or AAA - Lloyds DP notations for a redundant
DP system (class2 - AA) not including physical
failures of fire and flood. Which would need to be
AAA (class 3)

ABS – American Bureau of Shipping (American class
society)

AFI – Agreed for implementaton

ASK - Automatic Station Keeping

ASOG - Activity Specific Operational Guideline – a DST used to ensure the DP vessel

AUTR- AUTRO – DNV DP notations for a redundant DP system (class2) AUTR not including physical failures of fire and flood which would be AUTRO (class 3)

ATA – Automatic Thrust Assist - mooring notation for DNV Posmoor

AVR – Automatic Voltage Regulator

BSEE – Bureau of Safety and Environmental Enforcement

Billy Pugh – personel transfer 'basket'

BOR – Black Out Recovery

CAM – Critical Activity Mode

CEO – Chief Execuive Officer

CFO – Chief Financial Officer

COR – Center of Rotation

C of G – Center of Gravity

CPP – Controllable Pitch Propeller

CRT – Cathode Ray Tube

DG – Diesel Generator

DGNSS – Differential Global Navigation Satellite System

DNV – Det Norske Veritas (Norwegian class society)

DP – Dynamic Positioning

DP2 or DP3 ABS DP notations for a redundant DP system DP2 (class2) not including physical failures of fire and flood which would be DP3 (class 3)

DPCS - Dynamic Positioning control system

DPO – Dynamic Positioning Operator

DST - Decision Support Tool

DB - Derrick Barge (Crane Barge)

ECR – Engine Control Room

EDS - Emergency Disconnect System (disconnect from the well)

E Gen – Emergency Generator

ESD – Emergency Shutdown System

ET – Electronics Technician

FCC – Following Conference Call

FIR – Following Internal Review

FMEA – Failure Modes and Effects Analysis

FPP – Fixed Pitch Propeller

GEC - General Electric Company (aka Go Easy Crowd), later CEGELEC, Alstom, Converteam, GE

GMDSS – Global Marine Distress Signal Safety System

GPS – Global Positioning System

IGBT – Insulated Gate Bipolar Transistor

IJS – Independent Joy Stick

IMCA – International Marine Contractors
Association

IMO – International Maritime Organisation

INS – Inertial Navigation System

KM - Kongsberg Maritime

LIFE – Low Impact Failure Effect (design principle)

LOP – Loss of position

Lloyds – British Class Society

MDR – Marine Development Resource – term used
by Shell for DP consultants

MOC – Management of Change

MODU – Mobile Offshore Drilling Unit

MRT – Marine Risk Team (Shell)

MTS – Marine Technology Society

OSV – Offshore Support Vessel

PA – Public Address

PMS – Power Management System

PTW – Permit to Work

ROV – Remotely Operated Vehicle

SIMOPS – Simultaneous operations

Smoko – Aussie slang for tea/coffee break – on a ship this slang is also used for breaks that occur at 9 am and 3 pm in the engine control room (ECR) and bridge. If you want to find someone then it's best to go to the bridge or ECR. Things get discussed at Smoko, planned at Smoko, etc.

SOV – Statement of Verfication

STASCO – Shell Trading and Supply Company

TAL – Thruster Allocation Logic

TAMS – Thruster Assisted Mooring System

T's & C's – Terms and Conditions

UKOOA – United Kingdom Oil Operators Association

UPS – Uniterruptable Power Supply

USCG – United States Coast Guard

VFD - Variable Frequency Drive

VHF – Very High Frequency – broad term on ship that refers to the VHF radio communications

VOD - Vessel Overview Document

VRU – Vertical Reference Unit

WOW – waiting on weather

WSOG – Well Activity Specific Operational Guideline – a DST used to ensure the DP vessel, much like an ASOG but with points of disconnect from the well defined based on drift off rate and disconnect time.

BY THE SAME AUTHOR, ALSO AVAILABLE ON AMAZON:

POETIC LESSONS IN LIFE: Pithy Verse -
A Blend of Rhyme Maps, Poemedy and Epigrams

ANCHORS ARE FOR WANKERS

www.ingramcontent.com/pod-product-compliance
Lightning Source LLC
Chambersburg PA
CBHW071236220526
45468CB00002B/877